Contents

Preface

1	Characteristics and classification of living organisms	2
2	Organisation of the organism	7
3	Movement into and out of cells	11
4	Biological molecules	15
5	Enzymes	18
6	Plant nutrition	21
7	Human nutrition	27
8	Transport in plants	33
9	Transport in animals	36
10	Diseases and immunity	41
11	Gas exchange in humans	43
12	Respiration	48
13	Excretion in humans	51
14	Coordination and response	54
15	Drugs	59
16	Reproduction	61
17	Inheritance	68
18	Variation and selection	74
19	Organisms and their environment	78
20	Human influences on ecosystems	87
21	Biotechnology and genetic modification	91

Preface

This new edition of the *Cambridge IGCSE Biology Workbook* is designed as a 'write-in' book for students to practise and test their knowledge and understanding of the content of the Cambridge IGCSE™ Biology syllabus.

The sections are presented with the same headings and in the same order as in the Student's Book, *Cambridge IGCSE Biology Fourth Edition*, and as in the Cambridge IGCSE™ Biology syllabus for examination from 2023. All questions have been marked as either Core or Supplement and the relevant Student's Book chapter number matches the Workbook chapters. At the end of every section, there are longer questions (Exam-style questions), which aim to introduce students to an examination-style format. Note that these are not past paper questions, they are exam-style questions, written by the author of this book. The answers have also been written by the author. These questions are also identified as Core or Supplement.

To ensure the student's answers to the questions are kept together, there are spaces provided to write in. Extra paper may be needed to answer some questions.

This Workbook should be used as an additional resource throughout your Cambridge IGCSE Biology course alongside the Student's Book. The 'write-in' design is ideal for use in class by students or for homework.

1 Characteristics and classification of living organisms

Core

1 Draw lines to match each characteristic of living organisms to its definition.

Characteristic	Definition
nutrition	the chemical reactions in cells that break down nutrient molecules and release energy for metabolism
respiration	a permanent increase in size
excretion	making more of the same kind of organism
sensitivity	taking in materials for energy, growth and development
reproduction	an action by an organism or part of an organism causing a change of position or place
growth	the ability to detect and respond to changes in the internal or external environment
movement	the removal of waste products of metabolism

2 Distinguish between respiration and breathing, with reference to what happens in each process and where it occurs.

....

....

....

3 Explain why biologists do not accept defecation as an example of excretion.

....

....

4 Name **two** features that each of the following pairs of vertebrates have in common and **one** feature that makes them different.

Pair of vertebrates	Common feature		Feature that makes them different
fish and amphibians	1		
	2		
amphibians and reptiles	1		
	2		
birds and mammals	1		
	2		

5 Identify the invertebrate groups labelled A–D described on the invertebrate key. Record your answers below.

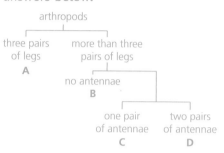

A ....

B ....

C ....

D ....

6 The diagram shows four species of flowering plants.

      A   B   C   D

Use the key below to identify the name of each flowering plant. Complete the table by putting a tick in the correct boxes to show how you have identified each plant. Plant A has been done for you.

1 **a** leaves narrow   go to 2

    b leaves broad   go to 3

2 **a** flowers bell-shaped   *Hyacinthoides non-scripta*

    b flowers trumpet-shaped   *Narcissus pseudonarcissus*

3 **a** leaves heart-shaped   *Ranunculus ficaria*

    b leaves club-shaped   *Primula vulgaris*

Plant	1a	1b	2a	2b	3a	3b	Name of plant
A	–	✓	–	–	–	✓	*Primula vulgaris*
B							
C							
D							

Supplement

7 a State **three** features that:

    i all flowering plants possess

        1 ....

        2 ....

        3 ....

    ii all flowering plant cells possess.

        1 ....

        2 ....

        3 ....

b Complete the table to distinguish between monocotyledons and dicotyledons.

Feature	Monocotyledon	Dicotyledon
leaf shape	long and narrow	
leaf veins		branching
cotyledons	one	
grouping of flower parts, e.g. petals		in fours or fives

8 Complete the table by naming each of the kingdoms of organisms described. The first has been done for you.

Kingdom	Description
animal	Multicellular organisms that have to obtain their food. Their cells do not have walls.
	Single-celled, with a nucleus. Some have chloroplasts.
	Many are made of hyphae, with nuclei and cell walls (containing chitin), but no chloroplasts.
	Multicellular organisms with the ability to make their own food through photosynthesis, due to the presence of chlorophyll. Their cells have walls (containing cellulose).
	Very small and single-celled, with cell walls, but no nucleus.

Exam-style questions

Core

1   A sheep is observed over a number of hours. State **three** observations that could be made to show that the sheep is a living organism.

    1   ....

    2   ....

    3   .... *[3]*

2   A car is non-living, but shows some characteristics that are also observed in living things.

    a   State **two** characteristics that the car has in common with living things.

    1   ....

    2   .... *[2]*

    b   State **two** characteristics of living things that a car does **not** show.

    1   ....

    2   .... *[2]*

                                                                                              [Total: 4]

3   The diagram shows a bacterium. It demonstrates a number of the characteristics of living things such as reproduction, movement and respiration.

    Name **three** other characteristics of living things that you would expect the bacterium to show.

    1   2   3   .... *[3]*

4 The diagram shows five species of an invertebrate group called molluscs.

Use the key to identify each species. Write your answers in the table.

molluscs

shell with two pieces —————————— shell with one piece

Cerastoderma edule

shell without coils ————————— shell coiled

Patella vulgata

oval opening to shell, coil flattened

shell opening with drawn out piece, coil long and pointed

shell opening small and round, very straight sided coil

Nucella lapillus

Littorina obtusata

Calliostoma ziziphinum

Letter	Species name
A	
B	
C	
D	
E	

[5]

Supplement

5 Explain how DNA is used in:

a the classification of organisms

....

....

.... [3]

b investigating the closeness of relationships between organisms.

....

....

.... [2]

[Total: 5]

2 Organisation of the organism

Core

1 State whether each of the statements is true or false. Circle T or F.

   a The outer layer of an animal cell is a cell wall. T / F

   b All cells have a nucleus. T / F

   c Some plant cells have chloroplasts for photosynthesis. T / F

   d Root cells do not have chloroplasts. T / F

   e Membranes are strong to stop the cell bursting. T / F

   f Cell walls are non-living. T / F

   g Cell walls are freely permeable, allowing water and salts to pass through. T / F

   h The vacuole of a cell contains cell structures such as chloroplasts. T / F

   i DNA is found in the nucleus. T / F

   j The cytoplasm of some white blood cells can flow to engulf microbes. T / F

2 a Label parts A–F on the diagram of the leaf palisade cell.

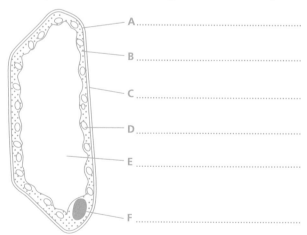

    A...

    B...

    C...

    D...

    E...

    F...

   b State **three** parts, present in this cell, which would not be found in an animal cell such as a liver cell.

    1 .... 2 .... 3 ....

3 Put the following terms in order of size from smallest to largest.

organ	nucleus	chromosome	cell	organ system	organism	tissue

(smallest)   (largest)

4 a State the formula for calculating magnification.

b A drawing of a grasshopper measures 12.0 cm from the end of its abdomen to its head, but the actual size of the organism is 2.8 cm. Calculate the magnification of the drawing. Show your working.

5 When viewed under a microscope at ×100, the apparent size of a protoctist is 15 mm. Calculate its actual size. Show your working.

6 a Define the term *tissue*. ....

....

b Complete the table by naming **two** types of animal cells and **two** types of plant cells and stating their functions.

Cell	Name	Function
animal cell 1		
animal cell 2		
plant cell 1		
plant cell 2		

7 State **two** functions for each of the following cell parts.

a cytoplasm

1 ....

2 ....

b cell membrane

1 ....

2 ....

   c nucleus

     1 ....

     2 ....

8 Some cells are specialised by having extensions with specific functions. Complete the table by naming **one** plant cell and **two** animal cells that are adapted in this way and stating their locations and functions.

	Name	Location	Function
plant cell			
animal cell 1			
animal cell 2			

Supplement

9 In a drawing of a liver cell, a student observed that the nucleus was 6 mm wide, which represented one-fifth of the width of the cell. The actual width of the cell was 60 µm.

Calculate the magnification of the drawing. Show your working at each step.

Exam-style questions

Core

1 a State **three** structural features that are present in both plant cells and animal cells.

     1 .... **2** .... **3** .... *[3]*

   b i Name **one** cell structure that would be found in a leaf palisade cell but not in a root hair cell.

     .... *[1]*

    ii State the chemical present in this cell structure and its function.

     Chemical ....

     Function .... *[2]*

                                                                                                          [Total: 6]

2 a Name **two** cell structures that distinguish plant cells from animal cells.

     1 ....

     2 .... *[2]*

b The diagram shows a group of animal cells. Complete the table by matching each of the functions described to a cell part **A–D**.

Function	Cell part
controls cell activities and development	
contains cell structures and is the site of chemical reactions	
wafts mucus and bacteria away from the lungs	
controls what substances enter and leave the cell	

[4]

[Total: 6]

3 Define the terms *organ*, *organ system* and *tissue*, naming **one** example of each in an animal and in a plant.

....

....

....

....

.... [6]

Supplement

4 The cells listed in the table below are adapted to their functions by having different numbers of some cell structures compared to other cells.

Complete the table for each cell:

a stating whether it has more, less or none of the named cell structure compared to typical cells and

b explaining how this adaptation benefits the cell's function.

Cell	Cell structure(s)	More/less/none	Explanation
red blood cell	nucleus		
upper epidermal cell	chloroplasts		
xylem vessel	nucleus		

[Total: 6]

5 Explain why a leaf is described as an organ.

....

....

.... [2]

3 Movement into and out of cells

Core

1 Define the term *diffusion*, making three key points. ....

....

....

2 In the lungs, a substance diffuses from the alveoli into the blood capillaries.

    a **i** Name the substance described. ....

       ii State **three** factors that will encourage a fast rate of diffusion for this substance.

         1 ....

         2 ....

         3 ....

    b Suggest and explain what would happen to the rate of diffusion if:

      i the temperature in the lungs went down ....

        ....

      ii the person's breathing rate increased. ....

        ....

3 Define the term *active transport*, making three key points. ....

....

....

Supplement

4 Two plants, **A** and **B**, were grown in soil with different nitrate ion concentrations. The concentration of nitrate ions was measured both in the soil and inside the roots of the plants. Both plants were found to be successfully absorbing the nitrate. The table shows the results.

Plant	Concentration of nitrate in the soil/arbitrary units	Concentration of nitrate in the roots/arbitrary units
A	157	156
B	82	114

a Complete the table below by stating the process or processes of absorption of nitrate ions in plants **A** and **B**. Give a reason for your choice.

Process(es)		Reason
plant A		
plant B		

b In another experiment, the roots of plant **B** were treated with cyanide (a respiratory poison). Explain why the movement of nitrates stopped in the plant.

....

....

5 Which process is responsible for the movement of the substances in the table below into or out of plant cells? Place a tick (✓) or a cross (✗) in each box to show your answer.

Substance	Diffusion	Osmosis	Active transport
oxygen			
water			
phosphate ions			
carbon dioxide			

6 Six identical potato cores were cut so that they all had the same mass and surface area. Each was placed in a sugar solution of a different concentration and left for 6 hours. They were then reweighed and the percentage change in mass was calculated. The graph shows the results.

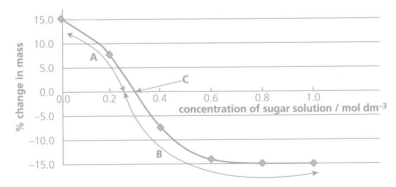

Describe and explain the results shown in zones **A**, **B** and **C** of the graph.

Zone **A** ....

....

....

Zone **B** ....

....

....

....

Zone **C** ....

....

....

Exam-style questions

Core

1 Which set of statements correctly describes active transport? Circle the letter of the correct set.

	Moves molecules from high concentration to low concentration	Moves molecules from low concentration to high concentration	Energy required	Energy not required
A	✓	✗	✓	✗
B	✗	✓	✗	✓
C	✓	✗	✗	✓
D	✗	✓	✓	✗

[1]

Supplement

2 Cells can obtain substances by diffusion, osmosis and active transport. Complete the table about the absorption of substances into two types of cell.

Type of cell	Substance absorbed	Process(es) used	Gradient: high to low / low to high?	Energy used?
root hair cell	water			
	nitrate	1 2	1 2	1 2
villus cell in small intestine	glucose	1 2	1 2	1 2

[15]

3 Six potato cores were all cut to the same length and then placed in a range of sugar solutions. After 1 hour, the cores were removed and re-measured. The results are shown in the table.

Sugar concentration/ mol dm⁻³	Start length/cm	Length after 1 hour/cm	Change in length/cm	% change in length
0.0	5.0	5.3		
0.2	5.0	5.1		
0.4	5.0	4.7		
0.6	5.0	4.4		
0.8	5.0	4.3		
1.0	5.0	4.3		

a Calculate the change in length of each potato core and write the results in the table. *[2]*

b Calculate the % change in length of each potato core and write the results in the table. *[2]*

c Plot a line graph of % change in length against concentration on the graph paper below. Use concentration as the *x*-axis.

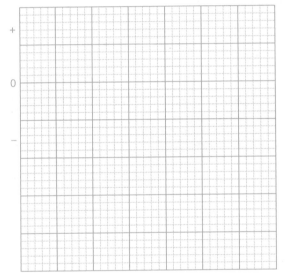

[4]

d i Use the graph to predict in which sugar concentration there would be no change in length of the potato core.

.... *[1]*

ii Explain why there would be no change in length at this concentration.

....

.... *[2]*

e Name **two** variables that should have been controlled in this investigation.

1 ....

2 .... *[2]*

f Suggest **one** way of making the results more reliable.

.... *[1]*

[Total: 14]

4 Biological molecules

Core

1 List the chemical elements present in:

a carbohydrates ....

b fats ....

c **all** proteins. ....

2 On the diagrams of food molecules (**a–c**):

i name each of the nutrients shown

ii state **one** use of the nutrient in the body

iii add labels to the diagrams, using words from the list below. You can use the words once or more than once.

> amino acid   fatty acid   chemical bond   glucose   glycerol

a ....

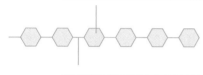

....

i ....

ii ....

....

b ....

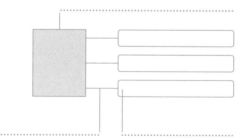

i ....

ii ....

....

....

c

i ....

ii ....

....

....

3 Draw lines to match the statements to the food tests and the results.

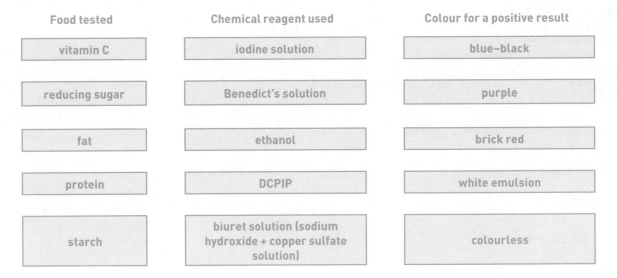

Food tested	Chemical reagent used	Colour for a positive result
vitamin C	iodine solution	blue–black
reducing sugar	Benedict's solution	purple
fat	ethanol	brick red
protein	DCPIP	white emulsion
starch	biuret solution (sodium hydroxide + copper sulfate solution)	colourless

Supplement

4 The diagram shows two strands of a section of DNA, with the bases on the left strand identified.

a Complete the diagram by writing the letters of the bases that would be present on the other strand.

b What name is given to describe the two strands coiled together?

c 28% of the bases in a DNA molecule are base **A**. Calculate the percentage of **C** bases. Show your working at each stage of the calculation.

Exam-style questions

Core

1 a Describe how you would carry out a food test on a piece of fish to find out if it contained oil.

....

.... *[3]*

b State what result would indicate a positive test for oil.

.... *[1]*

[Total: 4]

Supplement

2 Which two statements about DNA are correct? ....

1 Each strand contains chemicals called fatty acids.

2 T always pairs with A.

3 A gene is a length of DNA, coding for a protein.

4 A red blood cell contains DNA in its nucleus.

    A 1 and 3

    B 1 and 4

    C 2 and 3

    D 2 and 4 *[1]*

3 Outline the structure of DNA.

....

....

....

....

....

....

....

.... *[4]*

5 Enzymes

Core

1 Define the following terms, giving two key points for each term:

a catalyst ....

....

b enzymes. ....

....

2 The equation shows an enzyme-controlled reaction.

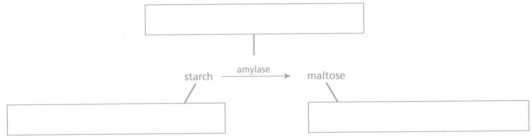

a Fill in the boxes to annotate the equation with the following terms:

enzyme   **end product**   **substrate**

b State **two** differences between a molecule such as starch (at the start of digestion) and glucose (an end product of digestion).

1 ....

2 ....

c Give **two** reasons why enzymes are important in all living organisms.

....

....

Supplement

3 a The graph shows the effect of temperature on the rate of an enzyme-controlled reaction.

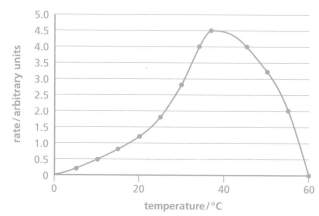

On the graph, use label lines to add the following labels:

denaturing is happening here

optimum reaction rate

the reaction is slow here (but could improve if conditions were improved)

the reaction is speeding up fastest here.

b Predict and explain what would happen if:

i the reaction mixture at 10 °C was warmed up to 30 °C ....

....

....

....

ii the reaction mixture at 60 °C was cooled down to 30 °C. ....

....

....

4 State **three** factors that affect the rate of enzyme-controlled reactions.

1 ....

2 ....

3 ....

Exam-style questions

Core

1 a Distinguish between a catalyst and an enzyme.

....

.... *[2]*

b Using appropriate terms, outline the action of an enzyme on the substance it breaks down.

....

....

.... *[3]*

[Total: 5]

Supplement

2 a Describe the action of a digestive enzyme on a large, insoluble food molecule.

....

....

....

....

.... *[5]*

b Explain why a digestive enzyme usually only works on one type of food molecule.

....

.... *[3]*

[Total: 8]

6 Plant nutrition

Core

1 a Write the **word equation** for photosynthesis.

b Name the molecule that glucose is converted to:

   i for transport around the plant ....

   ii for storage. ....

c Explain why chlorophyll is important in the process of photosynthesis.

   ....

d The diagram shows a section through a leaf. Complete the table by identifying parts **A–F** and
   stating their main functions.

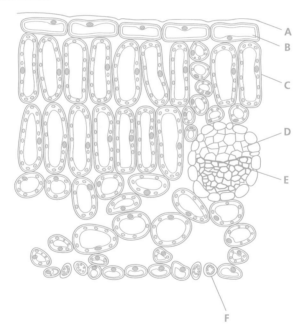

Leaf part	Name	Function
A		
B		
C		
D		

Leaf part	Name	Function
E		
F		

2 With reference to a leaf, name the cells that:

  a carry out most photosynthesis ....

  b control the exchange of gases such as CO_2, O_2 and water vapour ....

  c secrete a waxy cuticle for protection and waterproofing ....

  d transfer water into the leaf from the roots and stem ....

  e carry away the products of photosynthesis. ....

3 Describe how a leaf can be tested for starch. List **five** stages and give a reason for each stage used.

  ....

  ....

  ....

  ....

  ....

4 Explain why plants fail to grow well when there is a deficiency of:

  a nitrate ions ....

  ....

  b magnesium ions. ....

  ....

5 a The diagram shows the amount of light of different colours absorbed by chlorophyll.

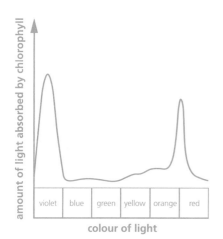

State which colours would be:

i   most useful to the plant ....

ii least useful to the plant. ....

b Some pond plants such as *Elodea* produce bubbles of oxygen when exposed to light. Describe how you could carry out an experiment to study the effect of light intensity on photosynthesis in *Elodea*. Make sure you detail how you will vary the light intensity, how you will control other variables, how many repeats you will carry out and any other precautions needed to carry out this investigation.

....

....

....

....

....

....

....

....

....

Supplement

6 Explain why plants grown in low-nitrate soils grow less well than plants grown in normal-nitrate soils.

....

....

7 Write a balanced chemical equation for photosynthesis.

8 Identify three limiting factors of photosynthesis. For each answer, state why it acts as a limiting factor.

1 ....

  Reason ....

2 ....

  Reason ....

3 ....

  Reason ....

Exam-style questions

Core

1 Outline an experiment which could be carried out to show that plants need carbon dioxide for photosynthesis.

....

....

....

....

....

.... *[8]*

2 The diagrams show cells found in a plant leaf.

    a **i**   On the diagrams, name each of the cells.

    A ....   C ....

    B ....      D ....

                                                          [4]

      ii Write the letters of the cells in the boxes below, in order from the upper surface of a leaf.

     □ ➡ □ ➡ □ ➡ □

                                                            [2]

    b Vascular bundles are also found in leaves. State **two** types of cell found in vascular bundles and outline their functions.

      Cell type 1: ....

      Function: ....

      Cell type 2: ....

      Function: .... *[6]*

                                                                 [Total: 12]

3 Which type of leaf cell has the most chloroplasts? Circle the letter of the correct answer.

    A upper epidermis   **C** spongy mesophyll

    B palisade mesophyll   **D** guard cell   *[1]*

Supplement

4 A student investigated the effect of light on the growth of pond plants over 8 weeks.

She set up three containers, each with the same volume of water and the same mass of plant.

Container **A** was lit by a 300 W bulb, **B** was lit by a 100 W bulb and **C** was lit by a 300 W bulb, with a coloured filter in front of the bulb.

Each week, the plants were taken out of the containers, carefully surface-dried, and weighed. Then they were put back into the containers.

The results are shown in the table.

Time / weeks	Mass / g		
	A	B	C
0	150	150	150
1	190	160	150
2	250	170	150
3	400	190	140
4	430	200	130
5	450	220	120
6	420	250	110
7	380	280	110
8	370	310	100

a Plot a graph of the results, labelling the axes and the line for each plant (**A**, **B** and **C**).

[6]

b Calculate the percentage increase in mass of the plant in container **A** during the first 5 weeks of the investigation.

[2]

c Suggest why the mass of the plants in container **A** began to decrease after week 5, while the mass of plants in container **B** continued to increase.

....

....

....

.... [2]

d During week 8, in which container would there be the least dissolved oxygen? Explain your answer.

....

....

.... [2]

[Total: 12]

7 Human nutrition

Core

1 Name the nutrient that is responsible for each of the following symptoms when in short supply.

   a soft bones and teeth ....

   b dehydration ....

   c constipation ....

   d scurvy ....

2 A teenager ate the same main meal each day: a hamburger in a white bread bun with chips and a glass of cola. The table shows the nutritional value of the food eaten.

Food	Quantity /g	Energy/ kJ	Protein /g	Animal fat/g	Carbohydrate /g	Calcium/ mg	Iron/mg	Vitamin C/ mg
hamburger	150	1560	30	15	30	50	4	0
bread	90	950	7	2	50	90	1	0
chips	200	2100	8	20	70	25	2	20
cola	300	550	0	0	30	0	0	0

   a With reference to the nutritional data, state **two** nutritional advantages and **two** disadvantages of eating this meal. Explain your answers.

   Advantage 1 ....

   Explanation ....

   Advantage 2 ....

   Explanation ....

   Disadvantage 1 ....

   Explanation ....

   Disadvantage 2 ....

   Explanation ....

   b Suggest **two** long-term effects of continuing to eat this meal regularly.

   1 ....

   2 ....

c This diet could be described as unbalanced.

    i Define the term *balanced diet*, making two key points. ....

    ....

    ii State **one** food group missing from the diet in the table. ....

3 On the diagram of the digestive system add labels to name parts **A–H**.

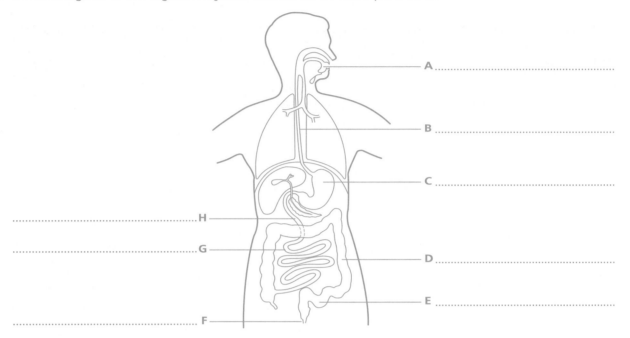

4 Complete the table to compare the types of human tooth.

Name	incisor		premolar	
Description		slightly more pointed than incisors		four/five cusps two/three roots
Function	biting off pieces of food		tearing and grinding food	

5 a Define the term *chemical digestion*, making two key points. ....

    ....

  b Complete the table by stating:

    i where, in the digestive system, each of the listed enzymes is secreted

    ii the substrates on which they act

iii the product(s) of digestion of these substrates.

Enzyme	Where the enzyme is secreted	The substrate on which the enzyme acts	The product(s) of digestion of the substrate
amylase	1 2		
lipase			1 2
protease	1 2		

Supplement

6 Outline the role of bile in the digestive system in

a physical digestion

b chemical digestion.

Make two key points for each type of digestion.

....

....

....

....

....

Exam-style questions

Core

1 Describe the functions of **two** named types of tooth in the physical breakdown of food.

1 type of tooth ....

function .... *[2]*

2 type of tooth ....

function .... *[2]*

[Total: 4]

2 a The diagram shows a section through a human tooth.

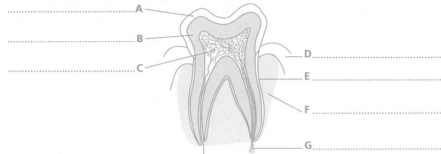

...................................A

...................................B

...................................C

D...................................

E...................................

F...................................

G...................................

     i On the diagram, label parts **A–G**.   *[7]*

    ii State the type of tooth shown and explain your answer.

      Type of tooth ....

      Explanation .... *[2]*

 b i Which layer of the tooth is the hardest? .... *[1]*

    ii Name **one** vitamin and **one** mineral needed to develop this layer.

      Vitamin ....

      Mineral .... *[2]*

[Total: 12]

3 A shortage of which food nutrient can result in the deficiency disease scurvy? Circle the letter of the correct nutrient.

  A vitamin C   **C** iron

  B vitamin D   **D** calcium   *[1]*

4 Enzymes are involved in the process of seed germination. Food stored in seeds is often in the form of starch.

  a i Name the enzyme needed to digest starch. .... *[1]*

    ii Name the product of this digestion. .... *[1]*

    iii Suggest **two** uses of this product in the germinating seed.

      1 ....

      2 .... *[2]*

    iv State **two** physical differences between starch and the product of its digestion.

      1 ....

      2 .... *[2]*

b The process of germination cannot start until the seed has taken in water.

Suggest **two** reasons why water is needed for germination.

1 ....

2 .... **[2]**

[Total: 8]

Supplement

5 The diagram shows a side view of a human abdomen.

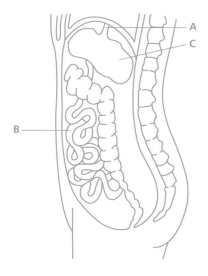

a Complete the table by identifying parts **A** and **B**. For each part, state **one** function.

Part	Name	Function
A		
B		

[4]

b State the type of food digested in organ **C**. .... **[1]**

[Total: 5]

6 Lamb kebab and Quorn® sausages are both good sources of protein. The nutritional content of these food products is shown in the table.

Food product	Nutritional content/100 g of product					
	Energy/kJ	Protein/g	Saturated fat/g	Fibre/g	Iron/mg	Calcium/mg
lamb kebab	1021	13.5	2.7	0.0	1.03	38.0
Quorn sausage	691	14.0	0.6	55	0.5	42.5

a   Using the data from the table, state and explain **two** reasons why Quorn sausages may be
    healthier than lamb kebab as a major item in the diet.

1   ....

2   .... [4]

b   Lamb kebab contains more iron than Quorn sausages.

    i   Suggest an additional source of iron for a person eating Quorn sausages in their diet.

        ....

        .... [1]

    ii State the function of iron in the body. ....

        .... [2]

    iii Outline the effects of a deficiency of iron. ....

        .... [2]

[Total: 9]

8 Transport in plants

Core

1 a State **two** functions of plant roots.

1 ....

2 ....

b Describe how root hair cells are adapted for their function, making two key points.

....

....

c The diagram shows a section through a root. List the parts through which water passes to get to the plant stem, putting them in the correct order. (Clue: only three parts are involved.)

....

2 The diagram shows a section through a young plant stem.

a Label the positions of the xylem and phloem on the diagram.

b State the functions of the xylem and phloem, making two key points for each.

Function of xylem: ....

Function of phloem: ....

Supplement

3 Describe how the structure of a xylem vessel is adapted to its function, making three key points.

....

....

4 Distinguish between the terms *transpiration* and *translocation*, making two key points for each.

....

....

....

....

5 State **two** factors that lead to an increase in the rate of transpiration.

1 ....   2 ....

<hr>

Exam-style questions

Core

1 What is the correct order of tissues through which water passes from soil to a leaf? Circle the letter of the correct answer.

  A root hairs → cortex → mesophyll   **C** root hairs → cortex → xylem → mesophyll

  B root hairs → cortex → phloem   **D** cortex → root hairs → xylem   *[1]*
    → mesophyll

2 The diagram shows an aphid feeding from phloem tubes in a plant stem.

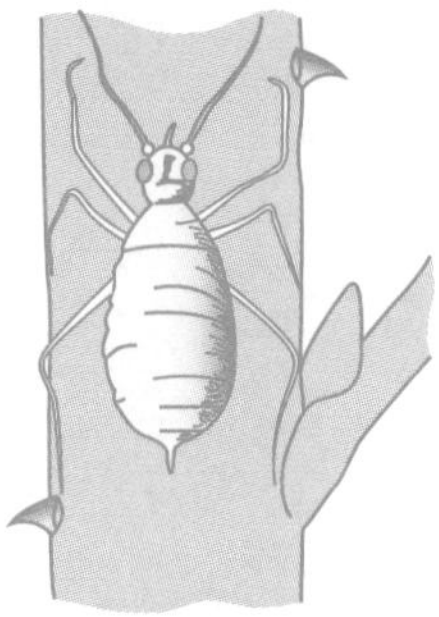

  What type of food would the aphid **not** receive? Circle the letter of the correct answer.

  A sugars   **B** amino acids   **C** water   **D** mineral ions *[1]*

3 Describe the process of water loss from a leaf.

....

....

....

....

....*[5]*

Supplement

4 A number of processes are involved in the movement of substances into and around a plant.
Complete the table by defining each term and explaining how it is important to a plant.

Process	Definition	Explanation
diffusion		
osmosis		
active transport		
transpiration		

[12]

5 a The diagram shows a section through part of a leaf. With reference to the features labelled
A–E, outline the process of water loss from the leaf.

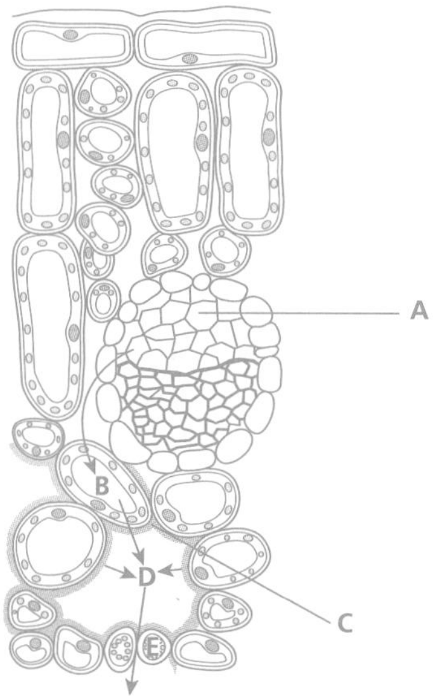

....

....

....

....

....

....

....

.... [6]

b Explain why wilting occurs.

....

....

.... [4]

[Total: 10]

9 Transport in animals

Core

1 **a** State which **two** heart chambers contain oxygenated blood.

1 ....   2 ....

b Name **three** different blood vessels, directly associated with the heart, which carry oxygenated blood.

1 ....   2 ....   3 ....

2 The table shows the pulse rate of an athlete before, during and after exercise.

Time/ min	0.0	1.0	1.5	2.0	2.5	3.0	3.5	4.0	4.5	5.0	5.5	6.0	6.5	7.0	7.5	8.0	8.5	9.0	9.5	10.0
Pulse rate/ beats per min	60	60	68	84	100	120	124	128	124	112	100	92	86	74	68	60	56	60	60	60

a Plot the data on the graph paper below, using time as the *x*-axis and pulse rate as the *y*-axis.

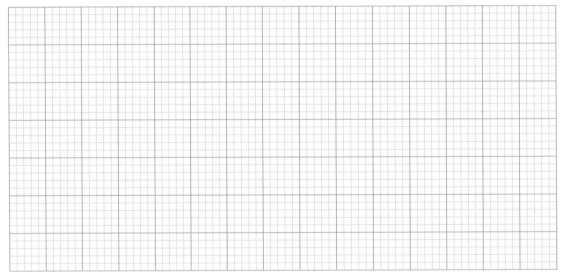

b **i** State the athlete's resting pulse rate. ....

ii Suggest when the exercise started. ....

iii At what time did the pulse rate first return to the resting pulse rate? ....

iv On the graph, sketch a line to represent the pulse rate of a less fit person doing the same exercise.

c Complete the table by identifying the main causes of heart attack and suggesting preventive measures. The first has been done for you.

Cause of heart attack	Suggestions for preventative measure
1 lack of exercise	start taking regular exercise
2	
3	
4	

Supplement

3 a Define the term *tissue*, making two key points. ....

....

b What type of tissue is the left ventricle made of? ....

c State **one** property of this tissue. ....

....

d Explain why the left ventricle has a thicker wall than the right ventricle, making three key points.

....

....

4 The diagram below shows the double circulatory system.

On the diagram:

a label the two boxes

b add arrows to the blood vessels and heart chambers to show the flow of blood

c shade the heart chambers and blood vessels that carry deoxygenated blood.

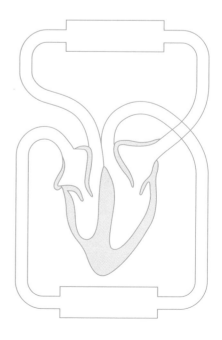

5 Explain briefly how the structure of a capillary is related to its function, making three key points.

....

....

....

6 a Describe **three** key features of the single circulation of a fish.

....

....

....

b Explain **two** advantages of a double circulation over a single circulation.

....

....

7 Explain why the pulse rate increases with exercise.

....

....

Exam-style questions

Core

1 The diagram shows a section through the heart.

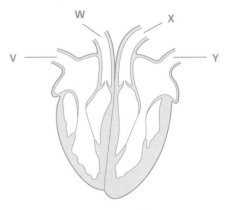

Which two blood vessels carry blood away from the heart? Circle the letter of the correct answer.

A V and Y   **B** W and X   **C** V and X   **D** W and Y   *[1]*

2 a Identify the parts of the heart labelled **A–K** on the diagram of the heart below.

.... K

.... J

.... I

.... H

A ....

B ....

C ....

D ....

E ....

F ....

G ....

[11]

b Name each of the parts of the heart described below:

i the chamber that receives deoxygenated blood from a vein ....

ii the blood vessel that passes blood from the lungs to the heart ....

iii the structure found at the start of the pulmonary artery ....

iv the heart chamber with the thickest wall ....

v the structure that separates the ventricles ....

vi the main vein of the body ....

vii the blood vessel that passes blood from the heart to the main body organs

....

viii the blood vessel which supplies the heart muscle with oxygen. .... *[8]*

[Total: 19]

3 a State **one** function for each of the following components of the blood.

i plasma ....

ii red blood cell ....

iii white blood cell ....

iv platelet .... *[4]*

b State **one** observable feature for each of the blood cells listed below.

i red blood cell ....

ii white blood cell .... *[2]*

[Total: 6]

Supplement

4 Which statement describes the structure and function of a lymphocyte? Circle the letter of the correct answer.

	Structure	Function
A	has no nucleus	transport of oxygen
B	has a nucleus	engulfs bacteria
C	has a nucleus	produces antibodies
D	has no nucleus	transport of water

[1]

5 a Describe the double circulation of a mammal.

....

....

....

.... [2]

b Complete the table by describing two distinguishing features of arteries, veins and capillaries.

Blood vessel	Distinguishing features
artery	1 2
vein	1 2
capillary	1 2

[6]

[Total: 8]

10 Diseases and immunity

Core

1 a Define the term *pathogen*. ....

b State **three** ways in which a pathogen can be transmitted.

1 ....

2 ....

3 ....

c Complete the table by giving **three** examples of body defences.

For each example, describe how it defends the body.

Body defence	How it defends the body
1	
2	
3	

Supplement

2 State what could be done to enhance the body's defences to protect a person from a life-threatening disease. ....

3 Outline the role of the immune system in:

a antibody production, making five key points ....

....

....

....

b phagocytosis, making three key points. ....

....

....

....

4 Explain the process of vaccination, making three key points. ....

....

....

....

5 Distinguish between passive immunity and active immunity. You should make four clear points.

....

....

....

Exam-style questions

Core

1 a Define the term *transmissible disease*.

....

.... *[2]*

b Describe how a pathogen is transmitted by

i direct contact ....

....

ii indirect contact. ....

.... *[2]*

[Total: 4]

Supplement

2 Explain the effect of cholera bacteria on the body.

....

....

....

....

....

.... *[6]*

11 Gas exchange in humans

Core

1 a Starting with the mouth, list the structures that air passes through in order for oxygen to be absorbed by a red blood cell.

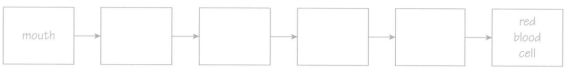

mouth → ⬜ → ⬜ → ⬜ → ⬜ → red blood cell

   b i Name the process by which oxygen passes across the wall of the alveolus into the

         bloodstream. ....

      ii List **four** features of a gas exchange surface that make it efficient.

          1 ....   3 ....

          2 ....   4 ....

2 The diagram shows structures in the human thorax.

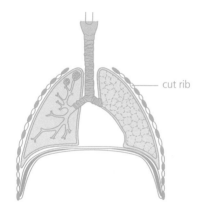

cut rib

   a Using label lines and the letters given below, label the following parts on the diagram.

      G where gas exchange happens

      A where air passes from the mouth to the bronchi

      M one place where muscles associated with breathing are found

   b Describe **two** effects of exercise on breathing. ....

      ....

  c The amount of carbon dioxide in expired air changes as a result of exercise.

     i Using data to support your answer, describe what happens to the amount of carbon dioxide in the expired air.

      ....

     ii Describe in three key points how you could test the expired air to see if it contains carbon dioxide.

      ....

      ....

3 The diagram shows a section through part of the lungs associated with gas exchange.

  a On the diagram, identify parts **A**, **B** and **C**.

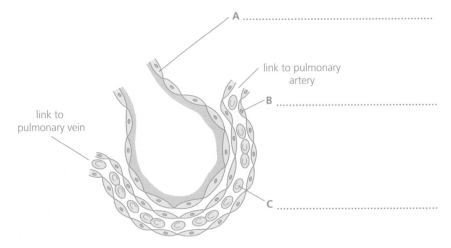

A ....

link to pulmonary artery

B ....

link to pulmonary vein

C ....

  b On the diagram, draw labelled arrows to show:

    i   the movement of oxygen

    ii the movement of carbon dioxide

    iii the direction of blood flow.

  c i By which process is oxygen passed into the blood? ....

    ii Explain **two** ways by which a concentration gradient is maintained to move oxygen into the blood.

      ....

      ....

Supplement

4 The following statements about breathing in (inspiration) are in the wrong order. Reorganise them to describe the process in the correct sequence.

air moves in to fill the lungs

air pressure in the lungs decreases

diaphragm moves down

diaphragm muscle contracts

external intercostal muscles contract

ribcage moves up and out

volume in the lungs increases

external intercostal muscles contract
↓
| |
↓
| |
↓
| |
↓
| |
↓
| |
↓
| |

5 a Describe **one** function of each of the following parts of the respiratory system.

1 cartilage in the trachea ....

2 ribcage ....

3 internal intercostal muscles ....

4 diaphragm ....

b Complete the table by explaining the roles of each of the cells and materials in the human gas exchange system.

Cell/material	Explanation of role
goblet cell	
ciliated cell	
mucus	

6 a The table compares the composition of inspired and expired air.

For each of the gases named, complete the table by explaining the difference between its percentage in inspired and expired air. The first one has been done for you.

Gas	Inspired air/%	Expired air/%	Explanation
nitrogen	79	79	*Not used or produced in body processes*
oxygen	21	16	
carbon dioxide	0.04	4	
water vapour	variable	saturated	

b During exercise, carbon dioxide builds up in muscle cells. Explain how this results in an increased rate of breathing. Your answer should include three key points.

....

....

Exam-style questions

Core

1 a Describe the effects of physical activity on breathing. ....

....

.... [4]

b i Describe an experiment to compare the amount of carbon dioxide in expired air before and after exercise.

....

....

....

....

.... [5]

ii State and explain the results you would expect from this experiment.

Results: ....

.... [2]

Explanation: ....

....

....

.... [3]

[Total: 14]

Supplement

2 The diagram shows a section through a human thorax (chest), including organs associated with breathing, gas exchange and circulation.

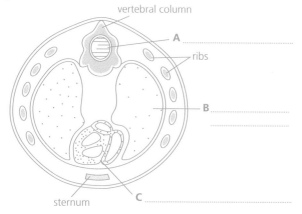

vertebral column

A ....

ribs

B ....

....

sternum

C ....

a On the diagram, identify the parts labelled **A**, **B** and **C**.   *[3]*

b i State which organ shown in the diagram has the largest surface area. .... *[1]*

 ii With reference to structures inside the organ, explain the importance of this large surface area.

....

....

.... *[4]*

c Outline the role of the ribs, intercostal muscles and diaphragm in breathing out (expiration).

....

....

....

....

....

.... *[6]*

[Total: 14]

12 Respiration

Core

1 a Write the **word equation** for aerobic respiration.

....

b State **three** uses of the energy produced in the bodies of humans.

1 ....   **3** ....

2 ....

c i Name the waste product of anaerobic respiration in muscles..

ii State why it is a disadvantage for an athlete to respire anaerobically.

....

....

2 When yeast respires anaerobically, two products are produced. Complete the table by naming the products and identifying a manufacturing process that makes use of each product.

Product	Manufacturing process
1	
2	

Supplement

3 When seeds germinate their rate of respiration increases, using glucose as a source of energy. Write a **balanced symbol equation** for aerobic respiration in the seeds.

....

Exam-style questions

Core

1 Which set of statements is correct? Circle the letter of the correct answer.

	Process	Carbon dioxide	Energy	
A	photosynthesis	produces carbon dioxide	traps energy	
B	photosynthesis	uses carbon dioxide	releases energy	
C	aerobic respiration	uses carbon dioxide	releases energy	
D	anaerobic respiration in muscles	produces no carbon dioxide	releases energy	[1]

2 A conical flask was set up with a mixture of yeast and sugar solution, as shown in the diagram.

The amount of bubbling was observed and the gas given off was passed through limewater. The time taken for the limewater to change colour was recorded. The experiment was carried out at five different temperatures. The results are shown in the table below.

Temperature/°C	Observation of bubbles	Time taken for limewater to change colour/s
0	no bubbles	no change
15	slow bubbling	320
30	many bubbles	65
45	slow bubbling	400
100	no bubbles	no change

a i Name the process carried out by the yeast to produce the bubbles of gas.

.... [1]

ii What gas is tested for using limewater? .... [1]

iii State the colour change of the limewater if this gas is present.

.... [1]

iv Name a chemical that would be present in the flask at the end of the experiment

at 30 °C that was **not** present at the start. .... [1]

b Outline the conclusions that can be drawn from the results of these experiments.

....

....

....

....

.... [5]

[Total: 9]

Supplement

3 a How do the products of anaerobic respiration in muscles and in yeast differ?

....

....

   b After vigorous exercise, an oxygen debt can build up. Explain:

   i why this happens ....

   .... [2]

   ii how the debt is removed. ....

   ....

   .... [3]

   [Total: 8]

13 Excretion in humans

Core

1 When urea is excreted, it passes through a number of structures before it is released as urine. Put the structures in the correct order, starting with the aorta.

> bladder   capillary   kidney   renal artery   ureter   urethra

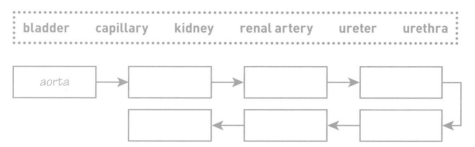

Supplement

2 Complete the paragraph about the function of the kidney, using words from the list.

> active uptake   excreted   blood   concentrated   diffusion   dilute   glucose
>
> less   more   osmosis   reabsorbed   water

The kidney filters ...., removing urea, excess .... and some ions.

All the .... is also filtered out, but it is all .... Some water is also

reabsorbed, depending on the state of hydration of the body. On a hot day ....

water is reabsorbed, resulting in a small amount of .... urine. Water is reabsorbed

by the process of ...., while glucose and ions are returned to the blood by

.... and ....

3 a Describe how the level of amino acids is controlled in the body, making three key points.

....

....

b State **two** functions of the liver, other than the control of amino acid levels.

1 ....

2 ....

4 The diagram shows a section through a kidney.

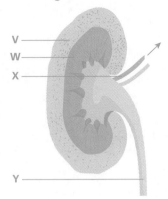

V ———
W———
X———

Y ———

a i Identify the labelled parts.

V .... X ....

W..................................... Y ....

ii State which of the parts you have named:

contains the glomeruli ....

transfers urine to the bladder. ....

b Outline the roles of the following parts of the kidney, making two key points for each:

i glomerulus ....

....

ii nephron ....

....

iii collecting duct. ....

....

Exam-style questions

Core

1 The diagram shows the urinary system.

a i On the diagram, label parts **A–E**. *[5]*

ii State **three** ways in which the substances *in solution* in part **A** would be different from those in part **C**.

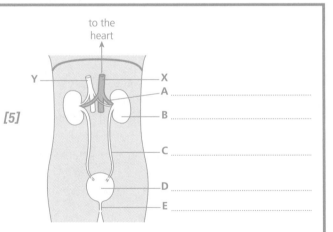

to the
heart

Y ———   ——— X
             ——— A ....
             ——— B ....

             ——— C ....

             ——— D ....
             ——— E ....

1 ....

2 ....

3 .... *[3]*

iii Suggest how the contents of part **D** would be different on a hot day compared to a cold day.

.... *[2]*

b State **three** differences between the structure of blood vessels **X** and **Y**.

1 ....

2 ....

3 .... *[3]*

[Total: 13]

Supplement

2 a Outline the structure and function of a nephron and its associated blood vessels.

....

....

....

....

....

....

.... *[6]*

b Explain why the kidney is an important organ in the body.

....

....

.... *[3]*

[Total: 9]

3 a State how amino acids travel from the small intestine to the liver.

....

.... *[2]*

b Describe how the liver assimilates amino acids.

....

.... *[1]*

c Outline the process of deamination.

....

.... *[2]*

[Total: 5]

14 Coordination and response

Core

1 a Distinguish between the central nervous system and the peripheral nervous system. Make four key points in your answer.

....

....

....

   b i Define the term *sense organ*. ....

....

     ii Complete the table by giving examples of sense organs and the stimuli they detect.

Sense organ	Stimulus detected
1 ear	
2	light
3 nose	
4	chemicals (taste)
5	temperature, pressure, touch, pain

2 State **three** differences between a sensory neurone and a motor neurone.

   1 ....

   2 ....

   3 ....

3 a Define the term *hormone,* making three key points.

....

....

   b A man experiences a sudden loud bang. This results in the secretion of adrenaline. Outline **three** effects of the adrenaline secretion on his body.

....

....

4 The diagram below shows a reflex arc involving a finger and arm.

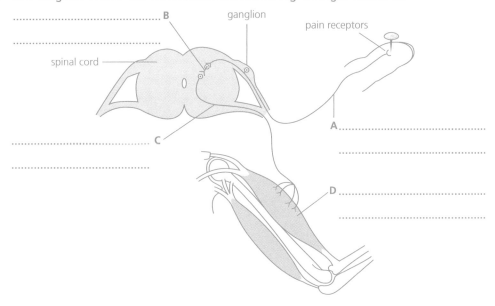

.... **B**

....

spinal cord ....

ganglion

pain receptors

A ....

....

.... **C**

....

D ....

....

a On the diagram:

   i identify neurones **A**, **B** and **C**

   ii identify the effector **D**

   iii add arrows to show the direction of the nerve impulse.

b State the effect on **D** of receiving the nerve impulse. ....

c Name the gap that links neurones **A** to **B**, and **B** to **C**. ....

5 a Define the term *gravitropism*. ....

....

b Complete the table:

   i by stating which part of a plant you would expect to grow towards gravity and which part would grow away from gravity

   ii by *suggesting* an advantage to the growth response.

	Advantage of the growth response
part growing towards gravity	
part growing away from gravity	

Supplement

6 The diagram below shows an eye exposed to bright light.

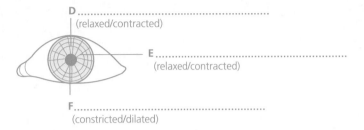

D ....
(relaxed/contracted)

E ....
(relaxed/contracted)

F ....
(constricted/dilated)

On the diagram:

a identify parts **D**, **E** and **F**

b complete the annotation by crossing out the incorrect statements.

7 Complete the table to compare the nervous and hormonal control systems.

Feature	Nervous	Hormonal (endocrine)
form of transmission		
transmission pathway		
speed of transmission		
duration of effect		

8 The diagram shows a section through part of the eye.

A

B

C

The parts labelled **A**, **B** and **C** are adjusted for distant vision.

In the table, identify parts **A**, **B** and **C** and describe what would happen to each part if the eye was exposed to a near object.

Name of part	Description of action
A	
B	
C	

9 a Name the chemical found in plants that controls phototropism. ....

  b i State the effect of shining a one-sided light onto a plant shoot.

....

    ii Explain how this effect is achieved in a shoot, making four key points.

....

....

....

Exam-style questions

Core

1 The diagram shows a cell.

  a What type of cell is shown? Circle the letter of the correct answer.

    A relay neurone   **C** sensory neurone

    B motor neurone   **D** cone cell   **[1]**

  b On the diagram:

    i label the following parts: cell body, dendrite   **[2]**

    ii place an **X** where a synapse could connect the neurone to another neurone.   **[1]**

[Total: 4]

Supplement

2 With reference to glucose levels in the blood, describe the role of negative feedback in homeostasis.

....

....

....

....

....

.... *[8]*

3 An athlete runs a race in hot conditions.

    a Describe the roles of blood vessels and sweat in reducing the body temperature of the athlete to normal.

        ....

        ....

        ....

        ....

        .... *[6]*

    b During the race, the athlete's muscles require more glucose to provide energy. Describe how the body provides the extra glucose.

        ....

        ....

        .... *[4]*

                                                                        [Total: 10]

4 A person suffering from Type 1 diabetes does not produce insulin when it is needed. As a result, the person needs regular injections of insulin.

    a What is the role of insulin in the body?

        ....

        .... *[2]*

    b When are levels of glucose in the blood most likely to be high?

        .... *[1]*

    c In addition to having regular injections of insulin, suggest two ways a person with Type 1 diabetes can manage the disease.

        ....

        .... *[2]*

                                                                           [Total: 5]

15 Drugs

Core

1 a Define the term *drug*. ....

....

b Which of the following can be described as a drug?

	Yes/no
A chemical used to fight an illness	
Aspirin – used to reduce the risk of a heart attack	
Paracetamol – used to reduce the pain of a headache	
Water – taken to reduce thirst	
Insulin – taken to reduce blood sugar levels	
Antibiotics – taken to stop a bacterial infection	
Blood transfusion – given to boost blood volume	

Supplement

2 a Describe how bacteria can become resistant to treatment by antibiotics.

....

....

b State **two** ways in which the development of resistant bacteria can be minimised.

1 ....

2 ....

Exam-style questions

Core

1 Describe how antibiotics are effective in treating a bacterial infection.

....

....

.... *[3]*

Supplement

2 The diagram shows an organism, **X**.

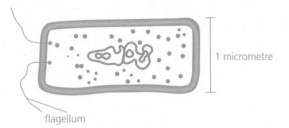

flagellum

1 micrometre

a i Identify the type of organism shown in the diagram.

.... *[1]*

ii State **three** reasons for your answer.

1 ....

2 ....

3 .... *[3]*

b The diagram shows how the reproduction of organism **X** is affected by an antibiotic.

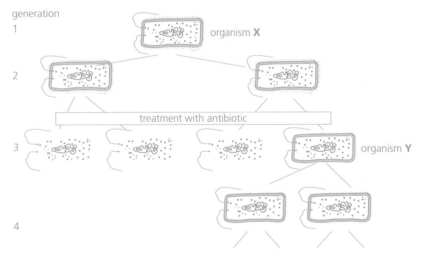

generation
1

organism **X**

2

treatment with antibiotic

3

organism **Y**

4

i State the type of reproduction shown by organism **X**. .... *[1]*

Organism Y has survived the antibiotic treatment but the other organisms in the third generation have been killed.

ii Suggest why organism **Y** and its offspring in the fourth generation have survived the antibiotic treatment.

....

.... *[2]*

[Total: 7]

16 Reproduction

Core

1 a Define the term *asexual reproduction*, making two key points.

....

....

b Distinguish between the terms *pollination* and *fertilisation*. Make two key points for each.

....

....

....

2 Draw lines to match each of the parts of the flower to its function.

Flower part		Function
anther		protects the flower while in bud
ovary		often large and coloured to attract insects
petal		produces pollen grains containing male sex cells
sepal		sticky, to receive pollen grains during pollination
stigma		contains ovules, with the female sex cells

3 The diagram shows a section through the reproductive parts of an insect-pollinated flower.

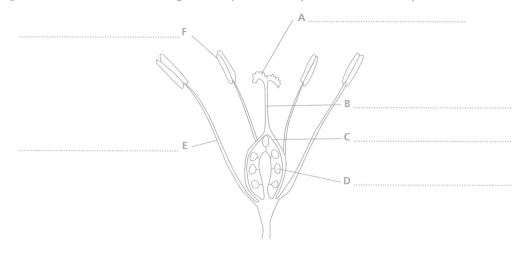

a On the diagram, label parts **A–F**.

b Name the reproductive structures made up of:

i parts **A**, **B** and **C** ....

ii parts **E** and **F**. ....

c Suggest and explain how the following would be different in a wind-pollinated flower:

i part **A** ....

....

ii the contents of part **F**. ....

....

4 Complete the crossword puzzle about the human reproductive system, using the clues to help you.

Across

5 adds fluid and nutrients to sperm, to form semen (8, 5)

7 where ova are produced (5)

Down

1 carries semen and urine through the penis (7)

2 receives the male penis during sexual intercourse (6)

3 male gonad that produces sperm (6)

4 urine is stored here (7)

6 tube carrying liquids (4)

5 **a** Outline the route taken by a sperm during and after sexual intercourse, from the testes to a fertile egg. Make six key points.

....

....

....

....

b Describe the process of fertilisation, making three key points.

....

....

....

6 Name each of the following, using their descriptions.

a The temporary structure that forms between the embryo and the lining of the uterus.

....

b The structures the baby passes through during birth.

1 ....

2 ....

c A fertilised egg, before it starts dividing. ....

d The liquid that acts as a shock absorber for the fetus. ....

e The process of shedding the uterus lining, along with blood. ....

f The site of fertilisation. ....

Supplement

7 The diagram shows the menstrual cycle, with the levels of two hormones and their effect on the lining of the uterus.

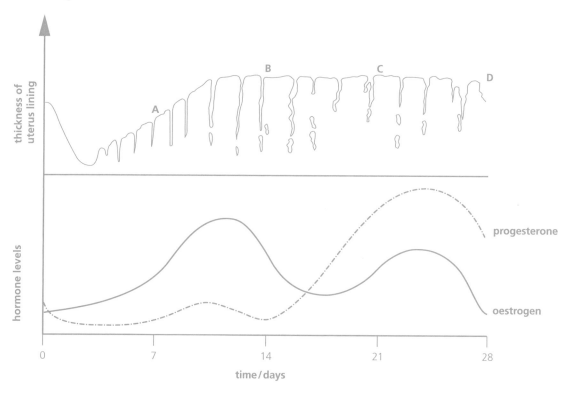

Describe what is happening at points **A**, **B**, **C** and **D**, referring to the causes of these events.

....

....

....

....

....

8 The diagram shows some apparatus that could be used to investigate the effect of temperature on the rate of respiration of germinating seeds.

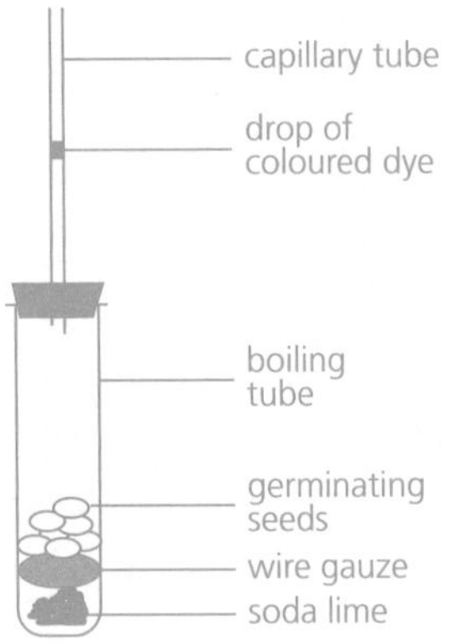

capillary tube

drop of coloured dye

boiling tube

germinating seeds

wire gauze

soda lime

a Describe how you would carry out this investigation, making six key points.

....

....

....

....

....

....

b Suggest how you would treat the results to show a relationship between temperature and the rate of respiration of the seeds.

....

....

....

Exam-style questions

Core

1 The diagram shows the structure of a flower.

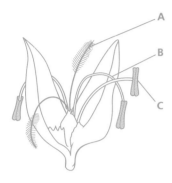

a State the type of pollination that this flower is adapted for. With reference to parts **A**, **B** and **C** explain your answer.

Type of pollination ....

Explanation ....

....

....

....

....*[7]*

b Outline how the process of pollination happens in this flower.

....

....

....*[4]*

[Total: 11]

2 The diagram shows the human male reproductive system.

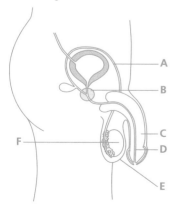

Complete the table by identifying the labelled structures and stating their functions.

Structure	Name of structure	Function
A		
B		
C		
D		1
		2
E		
F		

[13]

Supplement

3 Which of the following diseases is caused by a pathogen? Circle the letter of the correct answer.

A red–green colour blindness   **C** AIDS

B Type 1 diabetes   **D** coronary heart disease   [1]

4 The diagram shows a section through some of the reproductive parts of an insect-pollinated flower.

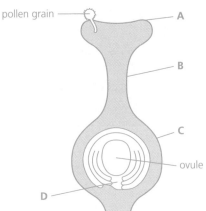

a i On the diagram, draw the route taken by the pollen grain to fertilise the ovule.   [3]

ii Describe how the pollen grain achieves fertilisation. Refer to the labelled parts in your answer.

....

....

....

....

....

....   [6]

b The diagram on the previous page represents part of an insect-pollinated flower. Suggest how the following parts would be different in a wind-pollinated flower.

   i the pollen grain .... *[1]*

   ii part **A** ....

     .... *[2]*

[Total: 12]

5 The diagram shows a front view of the reproductive organs of a woman.

a On the diagram, label parts **A**, **B**, **C**, **D** and **E**. *[5]*

b State and explain what feature, shown on the diagram, indicates that this woman is infertile.

Feature ....

Explanation .... *[2]*

c Name **three** structures which will form in the mother's reproductive system, so that the developing embryo will receive nutrients and be protected from physical harm.

   1 ....

   2 ....

   3 .... *[3]*

[Total: 10]

6 a Outline how cross-pollination can lead to variation in a species.

....

....

.... *[3]*

b Explain why a strawberry plant grown from a runner (stolon) will have identical characteristics to the plant which produced the runner.

....

.... *[2]*

[Total: 5]

17 Inheritance

Core

1 Draw lines to match each of the genetic terms to its definition.

Genetic term	Definition
allele	a length of DNA, coding for a protein
chromosome	having two identical alleles of a particular gene
dominant	a thread of DNA, made up of genes
gene	an allele that is always expressed if present
genotype	the observable features of an organism
homozygous	an alternative form of a gene
phenotype	the genetic make-up of an organism

2 Complete the diagram below to show how sex is inherited.

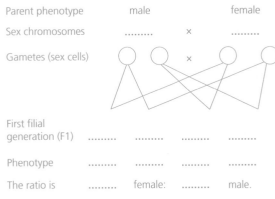

Parent phenotype   male   female

Sex chromosomes   ....   ×   ....

Gametes (sex cells)   ×

First filial
generation (F1)   ....

Phenotype   ....

The ratio is   ....   female: ....   male.

3 Fruit flies can have grey bodies or black bodies. Black body colour (**g**) is recessive to grey body colour (**G**). A pure-breeding grey fly was cross-bred with a pure-breeding black fly. The next (F1) generation were all grey bodied.

a Construct a genetic cross to show how the F1 generation were all grey bodied.

b Two flies from the F1 generation were cross-bred. Use a genetic cross to predict the genotypes and phenotypes of the next (F2) generation.

Supplement

4 a State **two** differences between the processes of mitosis and meiosis.

1 ....

2 ....

b A fruit fly has eight chromosomes in its body cells. Complete the table to show how many chromosomes would be present in the cells listed.

Type of cell	Number of chromosomes
leg muscle cell	
sperm cell	
zygote	
skin cell	

5 Snapdragon plants show codominance.

a Explain the term *codominance,* making three key points.

....

....

....

Pure-breeding snapdragon flowers can be red (C^RC^R) or white (C^WC^W).

b i Use a genetic cross to show how pink snapdragon flowers could be formed from pure-breeding parents.

ii Some of the pink-flowering snapdragon plants were self-pollinated. Predict the ratio of red, pink and white flowers that would be achieved from this cross.

....

Exam-style questions

Core

1 The diagram shows the results of some breeding experiments using rats.

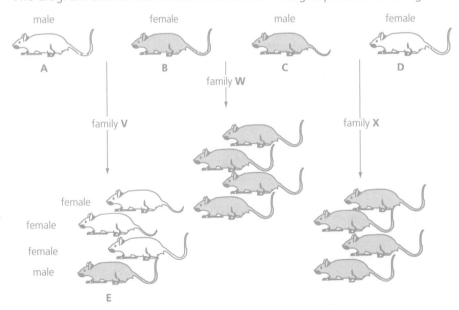

a i The inheritance of sex chromosomes is the same in rats as in humans. Complete the table to show the sex chromosomes present in the gametes of rats **B** and **C**.

Rat B		Rat C	

[1]

  ii If rats **B** and **C** had a second family, what is the percentage chance that the first rat born

would be male? ....   [1]

  iii Which sex cell determines the sex of the baby rat? Explain your answer.

....

....[2]

Coat colour in these rats is controlled by a single pair of alleles showing complete dominance.

b i Define the term *allele*. ....

....[2]

  ii Which of the parent mice **A–D** is likely to be:

    1 homozygous dominant for coat colour...........................   [1]

    2 heterozygous for coat colour? ....   [1]

c Grey rat **E** was later cross-bred with a grey rat of the same genotype as its mother (rat **B**). Write out a genetic cross to show how coat colour would be inherited in this family.

[5]

[Total: 13]

2 A man and a woman are both heterozygous for a condition called cystic fibrosis, but do not show any symptoms.

  a Choose suitable letters to represent the alleles. .... *[1]*

  b Draw a genetic diagram to show the chances of their children suffering from the condition.

[2]

  c Identify the genotypes and phenotypes of the children.

....

....*[2]*

[Total: 5]

Supplement

3 a Explain the term *sex-linked characteristic*. ....

....*[2]*

  b Explain why colour blindness is more common in men than in women.

....

....

....

....*[4]*

[Total: 6]

4 Use a genetic diagram to show how parents, neither of whom has blood group O, can have children with the blood groups O and AB. Use the symbols I^A, I^B and I^O to represent the alleles responsible for the human blood groups.

[5]

5 a Outline the process of mitosis (details of stages are not required).

....

....

.... *[4]*

b What are stem cells?

....

....

.... *[4]*

[Total: 8]

18 Variation and selection

Core

1 a Using examples, distinguish between continuous variation and discontinuous variation. In each case, state their causes. Your answer should include three key points for each type of variation.

....

....

....

....

....

....

   b Sketch graphs to illustrate the examples used in your answer.

2 a Define the term *adaptive feature*. ....

....

   b Using a named example, describe the process of natural selection, making five key points.

....

....

....

....

....

3 Describe how artificial selection can be used to produce a named variety of animal with increased economic importance. Make four key points in your answer.

....

....

....

....

Supplement

4 a i Define the term *gene mutation*. ....

....

ii State **two** causes of mutations.

1 ....

2 ....

b State **two** sources of genetic variation in populations, other than mutation.

1 ....

2 ....

5 The drawing shows a plant which is adapted to live in very dry conditions.

a What name is given to plants adapted to live in these conditions? ....

b Complete the table by identifying three **visible** adaptive features of the cactus shown and two features you would expect to find in a xerophyte which are **not** visible in the drawing.

For each answer, state how the adaptation helps the plant to survive.

	Visible adaptive feature	How the feature helps the plant to survive
1		
2		
3		

	Adaptive feature *not* visible	How the feature helps the plant to survive
1		
2		

Exam-style questions

Core

1 a Which of the following is an example of an adaptive feature? Circle the letter of the correct answer.

   A A girl with a pierced ear

   B A man with a suntan

   C The long neck of a giraffe

   D A polar bear with a long tail   *[1]*

   b Identify **three** adaptive features of the polar bear, which are visible on the diagram.

   1 ....

   2 ....

   3 .... *[3]*

[Total: 4]

Supplement

2 a State the type of variation shown by human blood groups. .... *[1]*

   b State an example of a different type of variation and explain how it is brought about.

     Example ....

     Explanation ....

     .... *[3]*

                                       [Total: 4]

3 a State the term used for plants which are modified to survive in conditions where they live in water.

     .... *[1]*

   b Name **one** plant that survives in very wet conditions and describe its modifications.

     Name ....

     Modifications ....

     .... *[3]*

                                       [Total: 4]

19 Organisms and their environment

Core

1 Draw lines to match each of the terms about food chains to its definition.

Term	Definition
carnivore	an animal that gets its energy by eating plants
consumer	an organism that makes its own organic nutrients, usually through photosynthesis
decomposer	an animal that gets its energy by eating other animals
food web	an organism that gets its energy by feeding on other organisms
herbivore	an organism that gets its energy from dead or waste organic material
producer	a network of interconnected food chains

2 State the **four** main processes involved in the carbon cycle.

1 ....

2 ....

3 ....

4 ....

3 a Define the term *population,* making three key points. ....

....

b i On the axes below, sketch a graph to show a curve of population growth, influenced by limiting factors.

ii On your graph, label the **four** phases of the growth curve.

c State **three** factors that limit population growth.

1 ....

2 ....

3 ....

d Describe the effect of a lack of limiting factors on a graph of human population growth. Make two key points in your answer.

....

....

4 The diagram below shows a food web.

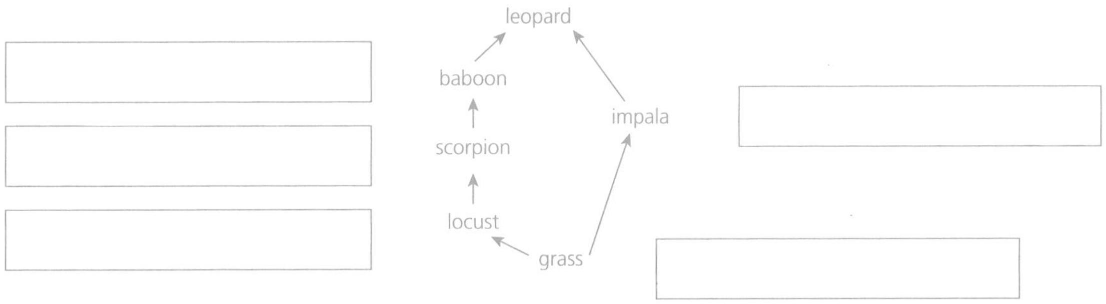

leopard

baboon

scorpion

locust

impala

grass

a In the boxes provided, state the trophic level of each of the organisms, other than the leopard.

b Suggest why the trophic level of the leopard is difficult to label in this food web.

....

c Food webs can easily become unbalanced by the death of the population of one organism. Suggest **three** causes of the death of a population.

1 ....

2 ....

3 ....

d The organisms in the food web are linked by arrows. Explain what the arrows represent.

....

Supplement

5 A bird of prey called a sparrow hawk hunts in a field of clover plants, searching for its prey: another bird called a thrush. The thrushes visit the field to feed on snails. The snails graze on the clover leaves.

a Explain the meaning of the term *food chain*, making two key points.

....

....

b i Draw a food chain using the organisms described in the paragraph.

ii Add the following labels to the names of the organisms in the food chain.

> **primary consumer**   **producer**   **tertiary consumer**   **secondary consumer**

c A farmer sprayed the clover field with a molluscicide, killing all the snails. Suggest and explain the effects of this treatment on:

i thrushes ....

ii sparrow hawks. ....

d Suggest **two** reasons why the percentage energy lost at the snail trophic level of the food chain is less than that lost at the thrush trophic level.

....

....

6 a Plants form the first trophic level of food chains.

i State the process used by plants to obtain their energy. ....

ii What is the source of this energy? ....

iii Give **three** reasons why plants do not make use of all the energy available to them.

1 ....

2 ....

3 ....

b **i** Explain why there tend to be small numbers of top carnivores in a food chain.

....

....

ii Explain why short food chains are more efficient than long food chains.

....

....

7 The diagram shows an inverted pyramid of numbers.

```
            ┌─────────────────────────┐
            │          fleas           │
            └─────────────────────────┘
                │      owl      │
            ┌───────────────────────┐
            │       blue tits        │
        ┌───────────────────────────────┐
        │          caterpillars          │
        └───────────────────────────────┘
                │   oak    │
                │   tree   │
```

a Explain why pyramids of numbers, like the one shown above, can be inverted. Make two key points.

....

....

b Explain why pyramids of biomass nearly always have a normal pyramid shape.

....

c Suggest why the process of collecting data for pyramids of biomass is destructive.

....

....

8 **a** Describe the roles of each of the following in the nitrogen cycle. Make two key points for each.

i Nitrogen-fixing bacteria ....

....

ii Nitrifying bacteria ....

....

iii Denitrifying bacteria ....

....

  b State **three** strategies that farmers can use to improve the nitrates in soil.

    1 ....

    2 ....

    3 ....

9 There is a delicate balance between the amounts of carbon dioxide and oxygen in the atmosphere.

  a State the **two** processes, occurring in living things, which affect this balance.

    1 ....   2 ....

  b Name **two** human activities which, when carried out on a large scale, can affect this balance.

    1 ....   2 ....

Exam-style questions

Core

1 Perch are fish which feed on smaller fish called minnows. Minnows feed on crustaceans called water fleas. The water fleas eat microscopic algae.

  a **i** Draw a food chain using the organisms named in the information above.

    [2]

    ii Add the following labels under the names of the organisms in the food chain.

       tertiary consumer   secondary consumer   producer   primary consumer *[2]*

  b When the organisms in the food chain die, their bodies are broken down by other organisms such as bacteria and fungi. State the name used to describe organisms that feed in this way.

    .... *[1]*

    [Total: 5]

2   The table shows the biomass of living organisms in a marine pool.

Organism	Position in food chain	Biomass/g per square metre
seaweed	producer	18.20
periwinkle	primary consumer	1.32
crab	secondary consumer	0.55
fish	tertiary consumer	0.12

a   State the term used to describe an organism's position in a food chain.

....

[1]

b   The crab is a secondary consumer. With reference to organisms in the table, explain what the term *secondary consumer* means.

....

.... [2]

c   Draw a pyramid of biomass for the data in the table. Add labels to identify the levels.   [3]

d   Describe what happens to the biomass as you move up a food chain.

.... [1]

e   The relationship between the organisms in the pyramid of biomass could be displayed as a food chain, with arrows linking the organisms.

What do the arrows represent in a food chain?

....

.... [2]

[Total: 9]

Supplement

3 a i Define the term *trophic level.* ....
.... *[1]*

ii Draw an example of a food chain involving **four** trophic levels.

[2]

iii Draw and label a pyramid of biomass for your food chain.

[2]

b Plants obtain their energy from sunlight. 60% of this light is the wrong wavelength, or is reflected off the surface of the leaf. Another 5% passes through the leaf without being captured.

i Calculate the percentage of the light available that is captured by the leaf.

.... *[1]*

ii What feature is present on the surface of a leaf that could reflect the light?

.... *[1]*

iii The main tissue in a leaf responsible for capturing light is the palisade layer. Describe how cells in this layer are adapted for efficient light absorption.

....
.... *[2]*

c As energy passes through the animals in the food chain, some is lost.

i State what percentage of energy is lost between trophic levels. .... *[1]*

ii State **two** ways in which animals use this energy.

1 ....

2 .... *[2]*

[Total: 12]

4 The diagram shows the energy in two food chains, from which humans obtain their energy.

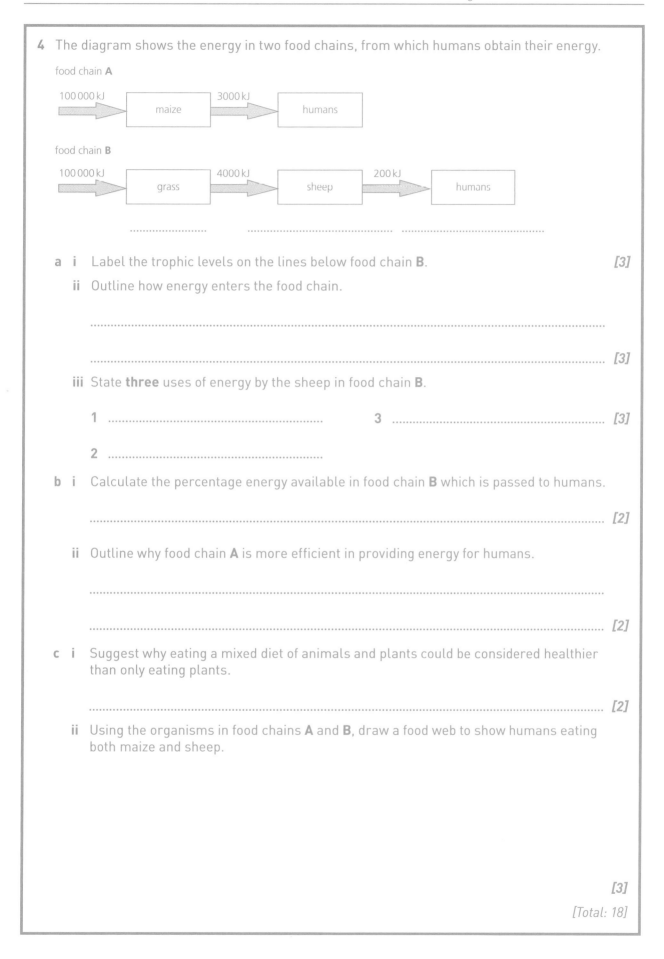

food chain **A**

100 000 kJ → maize → 3000 kJ → humans

food chain **B**

100 000 kJ → grass → 4000 kJ → sheep → 200 kJ → humans

....

a i   Label the trophic levels on the lines below food chain **B**.   *[3]*

   ii Outline how energy enters the food chain.

       ....

       .... *[3]*

   iii State **three** uses of energy by the sheep in food chain **B**.

       1 ....   3 .... *[3]*

       2 ....

b i   Calculate the percentage energy available in food chain **B** which is passed to humans.

       .... *[2]*

   ii Outline why food chain **A** is more efficient in providing energy for humans.

       ....

       .... *[2]*

c i   Suggest why eating a mixed diet of animals and plants could be considered healthier
       than only eating plants.

       .... *[2]*

   ii Using the organisms in food chains **A** and **B**, draw a food web to show humans eating
       both maize and sheep.

       [3]

       [Total: 18]

5 The diagram shows a bean, which is a leguminous plant, and maize, which is a non-leguminous plant. Both are growing in the same soil in a farmer's field.

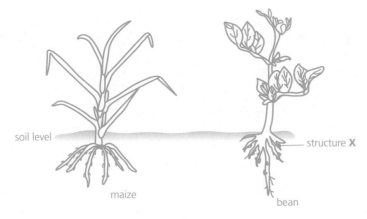

soil level

structure **X**

maize

bean

a i Identify structure **X.** .... *[1]*

ii Structure **X** contains organisms. State the group to which they belong.

.... *[1]*

iii The bean plant grew better than the maize plant. The farmer tested the soil and found it was low in nitrates. Explain why the bean plant grew well, even though the nitrate levels were low.

....

....

.... *[4]*

b Maize is a monocotyledon and the bean is a dicotyledon. Complete the table by identifying **two** features that help to classify these plants.

Feature	Description of feature for	
	Maize	Bean
1		
2		

[3]

[Total: 9]

Human influences on ecosystems

Core

1 Outline **four** ways in which modern technology has been used to increase food production.

1 ....

....

2 ....

....

3 ....

....

4 ....

....

2 Describe the effects of untreated sewage on a water ecosystem. Make five key points.

....

....

....

....

....

3 A logging company applied for permission to cut down a large area of forest. An environmental campaign group objected to the plan.

a State the term used to describe the removal of large areas of forest. ....

b Suggest **three** reasons why the logging company might have been planning to cut down the forest.

1 ....

2 ....

3 ....

c The environmental campaign group made a list of the undesirable effects on the environment of cutting down the trees and explained their possible consequences. Describe and explain **four** possible effects.

1 ....

....

2 ....

....

3 ....

....

4 ....

....

Supplement

4 The following statements describe the process of eutrophication in a lake, caused by water pollution by fertilisers. They are in the wrong order. Reorganise them in the correct sequence, writing the letter for each stage in the boxes.

A algae absorb fertiliser and grow rapidly (algal bloom)

B algae die without light

C fertilisers are very soluble and are easily leached out of the soil

D animals in water die through lack of oxygen

E algae form a blanket on the surface of the water, blocking light from the algae below

F bacteria decompose dead algae, using up oxygen in the water for respiration

G fertilisers are washed into the lake

C → ☐ → ☐ → ☐ → ☐ → ☐ → D

5 Describe **four** methods of conserving fish stocks.

....

....

....

....

6 Outline **three** reasons why global warming is considered to be an environmental problem.

....

....

Exam-style questions

Core

1 Outline how each of the applications of modern technology listed in the table has helped to increase food production. Also give **one** possible disadvantage of each application. Write your responses in the table.

Application	How it helps to increase food production	Possible disadvantage of its use
chemical fertiliser		
insecticide		
herbicide		

[9]

2 a Using an example familiar to you, describe the importance of conserving a species of animal that is under threat of extinction.

....

....

.... [4]

b For the species you have named:

 i describe the habitat in which it lives

....

.... [2]

 ii outline how this habitat could be conserved to protect the species.

....

....

.... [3]

[Total: 9]

Supplement

3 The following statements relate to the effects of applying too much nitrogen fertiliser, but they are in the wrong order. Place them in the correct sequence in the blank boxes.

> **decay by bacteria**   **rapid algal growth**   **leaching**
>
> **death of aquatic animals**   **death of algae**

☐ → ☐ → ☐ → ☐ → ☐

[2]

4 a Define the term *sustainable resource*. ....

.... *[2]*

b Outline **three** ways in which resources such as forests can be conserved.

1 ....

....

2 ....

....

3 ....

.... *[6]*

[Total: 8]

21 Biotechnology and genetic modification

Core

1 a State **two** features of bacteria that make them useful in biotechnology and genetic modification.

1 ....

2 ....

b Describe how yeast can be used to make biofuels. Make three key points.

....

....

....

c i Name the group of molecules to which pectinase belongs. ....

ii State what properties these molecules have that make them useful in biotechnology.

....

....

iii Describe the use of pectinase in fruit-juice production. Make four key points.

....

....

....

Supplement

2 Name **two** industries that use enzymes in making foods and describe the role of enzymes in these processes.

Industry 1 ....

Description ....

Industry 2 ....

Description ....

3 a The diagram shows a fermenter, used to manufacture enzymes.

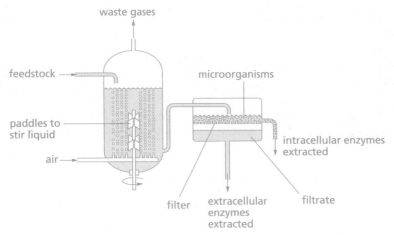

Describe the role of each of the following parts:

i paddles ....

ii filter ....

iii feedstock. ....

b What **two** types of organism are used to manufacture enzymes?

....

c Describe how intracellular enzymes are extracted from microorganisms, making two key points.

....

....

4 State **two** advantages and **two** disadvantages of genetically modifying a crop such as soya, maize or rice.

Advantages 1 ....

2 ....

Disadvantages 1 ....

2 ....

Exam-style questions

Core

1 a Define the term *genetic modification*.

....

.... *[3]*

   b State **three** examples of genetic modification.

   1 ....

   2 ....

   3 .... *[3]*

[Total: 6]

Supplement

2 Describe how genetic modification can be used to make a human protein such as insulin.

....

....

....

....

....

....

.... *[8]*

Reinforce learning and deepen understanding of the key concepts in this Workbook, which provides additional support for the accompanying Cambridge IGCSE™ Biology Student's Book.

» **Differentiated content:** both Core and Supplement content is clearly flagged with differentiated questions testing content.

» **Provide extra practice and self-assessment:** each Workbook is intended to be used by students for practice and homework. Once completed, it can be kept and used for revision.

» **Develop understanding and build confidence ahead of assessment:** the Workbook is in syllabus order, topic-by-topic, with each section containing a range of shorter questions to test knowledge, and sections providing exam-style questions.

Use with *Cambridge IGCSE™ Biology Student's Book Fourth Edition*
9781398310452

For over 30 years we have been trusted by Cambridge schools around the world to provide quality support for teaching and learning. For this reason we have been selected by Cambridge Assessment International Education as an official publisher of endorsed material for their syllabuses.

This resource is endorsed by Cambridge Assessment International Education

✓ Provides learner support for the Cambridge IGCSE and IGCSE (9-1) Biology syllabuses (0610/0970) for examination from 2023

✓ Has passed Cambridge International's rigorous quality-assurance process

✓ Developed by subject experts

✓ For Cambridge schools worldwide

HODDER EDUCATION
e: education@hachette.co.uk
w: hoddereducation.com

ISBN 978-1-398-31049-0

MIX
Paper | Supporting responsible forestry
FSC™ C104740